The Earthworm and The Cockroach

nbillett

Published by nbillett, 2026.

THE EARTHWORM AND THE COCKROACH

First edition. February 25, 2026.

ISBN: 979-8233766107

Written by nbillett.

Table of Contents

Habits of the Earthworm 1
The Importance of the Earthworm to Man 2
The External Features of the Earthworm 3
Movement 5
The Internal Structure of the Earthworm 6
The Body Wall 7
The Alimentary Canal 8
The Blood System 10
The Excretory Organs 11
The Nervous System 12
Respiration 13
The Reproduction of the Earthworm 14
The Cockroach 17
Internal Structure of a Cockroach 22
Life History and Habits 26

Habits of the Earthworm

Earthworms are to be found in the surface layers of every moist soil and are especially numerous if the soil is rich in humus. If the surface soil becomes too dry, they descend to moister regions. They spend the daytime in the ground, but at night, especially if the surface is damp, they come out and crawl about in search of food. Frequently a worm does not come right out of its burrow but keeps the end of its body in the hole so that it can retire underground quickly if danger threatens.

The worm tunnels are partly by force and partly by eating its way through the soil. It is able to obtain nourishment from the small proportion of decayed animal and plant remains that are present in the soil which is swallowed. The soil is ground up by a part of the food canal known as the gizzard, and the finely powdered soil that is passed out of the body form the little heaps at the entrance to the burrow, which are known as worm casts. The burrow is lined with a layer of this fine soil, stuck together by the fluid secreted by the worm's skin.

A worm has no eyes but can tell when it is near the surface, for its front end is sensitive to bright light, which the worm avoids. A worm has no ears either, but it is able to perceive vibrations in the soil, and so will not venture from its burrow while a bird is hopping about overhead.

The Importance of the Earthworm to Man

Few people realize what an important part earthworms play in relation to agriculture. By their burrowing activities worms loosen the soil so that the roots of plants can grow down more easily. In addition, the entrance of rain and air into the soil is helped by the burrows, which both drain and aerate the soil. Leaves which have been dragged into burrows by worms, and which have not been eaten, enrich the humus content of the soil when they decay.

Soil with earthworm

The External Features of the Earthworm

The body of the worm is roughly cylindrical in shape, except at the hinder end, which is flattened from above downwards, and it is marked by transverse grooves into about 150 rings or segments. The long-pointed shape of the body, which may be 6 inches long, aids the worm in making its way through the ground. The dorsal surface is darker than the ventral surface, and the whole body is covered by a thin layer of cuticle, through which a blood vessel can be seen running along the dorsal surface. The surface of the worm is moist to the touch owing to a secretion from tiny glands, which is poisonous and so keeps the skin clear from bacteria and fungi. The secretion acts as a lubricant during the worm's movement through the soil and helps to stick the soil particles together in the wall of the burrow. It also maintains the skin in the moist condition that is essential for the respiration of the worm.

On the dorsal surface of the worm are small holes, the dorsal pores, which open in the grooves between the segments and place the interior of the body in communication with the exterior. One dorsal pore occurs in each groove behind the 10th segment. The first segment is called the peristomium and surrounds the mouth, which is overhung by a flap known as the prostomium. The anus is a small opening in the last segment of the body. About a quarter of the way along the body from the anterior end are five or six segments which have a swollen appearance. This is a glandular region called the clitellum, which secretes the cocoon into which the worm lays its eggs.

If a worm is drawn through the fingers a roughness due to small bristles can be felt. These bristles, which are called setae, project from the body wall in two double rows on both sides of the body, so that every segment, except the first and last, bears four pairs of setae. On each side one pair of setae occurs laterally, and the other pair lies beneath it nearer the ventral surface.

The only easily visible openings of the various internal organs are those of the male reproductive organs, which occur on the lower side of the 15th segment. They are transverse slits surrounded by fleshly lips.

Movement

When the worm is crawling over the surface the setae in the hinder part of the body are pushed into the ground and the circular muscles in the front part of the body contract; the pressure of the body fluid so set up then causes the front part of the worm to mov forwards. Next the setae of the front part of the body are pushed into the ground and those of the hind part are withdrawn. Finally, by the contraction of the longitudinal muscles and the relaxation of the circular muscles, the hind part is drawn into a new position. Thus, by the aid of the setae the worm moves over the ground by the alternate shortening and lengthening of the body. It is due to the projection of the setae that it is easier to break a worm in two than to draw it out of its burrow.

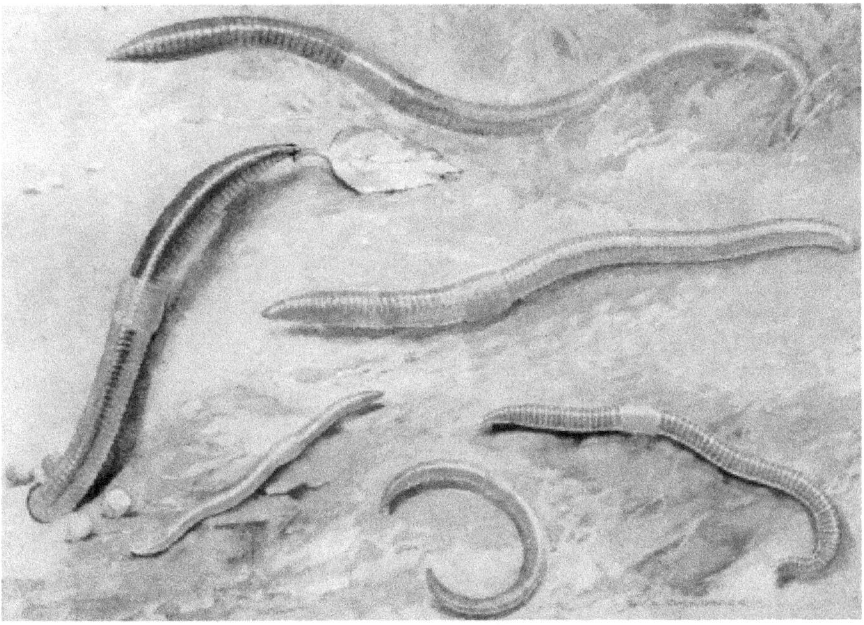

Worm diversity

The Internal Structure of the Earthworm

The mesoderm forms two layers one cell thick called the peritoneum. One layer adheres to the ectoderm and one of the endoderm. The space between them is called the coelom and contains the coelomic fluids. Animals which possess a coelom are referred to as the Coelomata.

The main structure of the worm consists essentially of two tubes, one fitted inside the other. The inner tube is the food tube or alimentary canal, and the other one is the body wall of the worm. The space between the two tubes is the coelom, and in the worm, it is divided into a number of chambers, each corresponding to a segment, by transverse partitions called septa. The middle of the lower edge of each septum is not fixed to the body wall, and the gap thus left allows the coelomic fluid to pass from one segment to another.

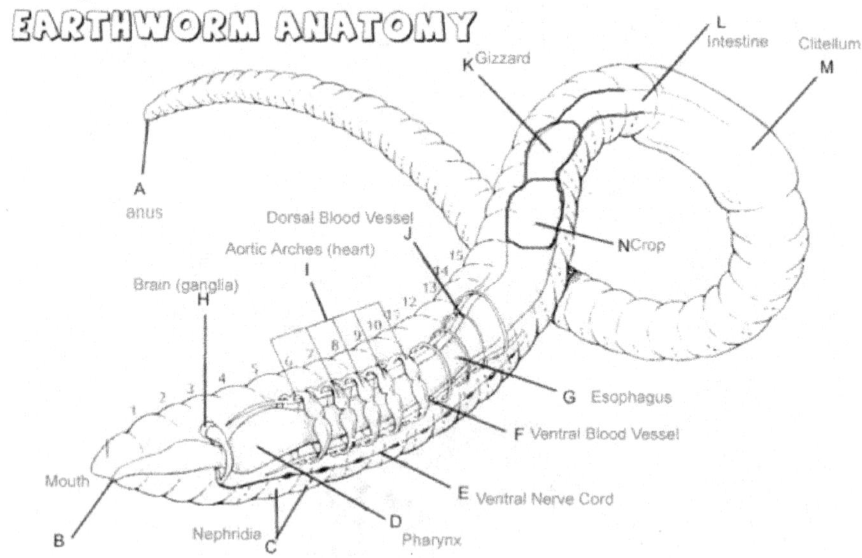

The Body Wall

The surface of the body is covered by a thin membrane, the cuticle, which is secreted by the outermost layer of cells, or epidermis. The epidermis corresponds to the ectoderm of Hydra. Beneath the epidermis lie two layers of muscle, an outer layer in which the fibers are arranged circularly round the body and an inner layer in which they run longitudinally. The setae are lodged in small sacs, which are inrushings of the epidermis.

Organs of an Earth Worm

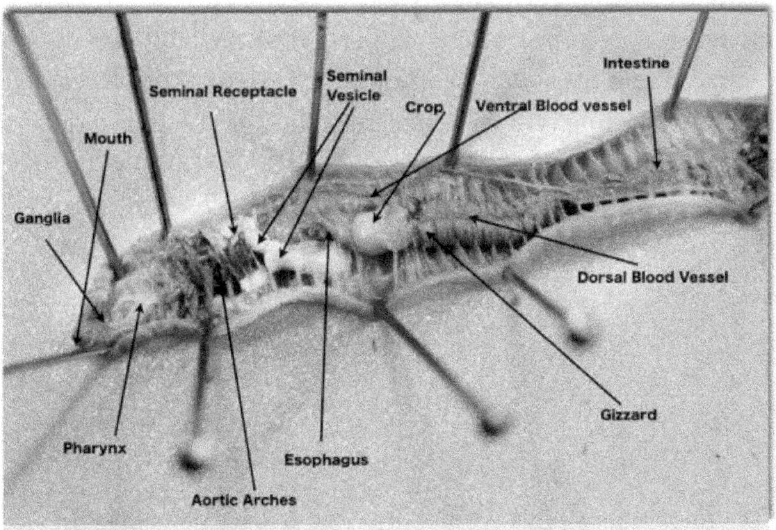

The Alimentary Canal

The mouth is a small crescent-shaped opening bordered by the peristomium and is overhung by the prostomium. It leads into a small chamber, the buccal cavity, which is followed by a larger chamber, the pharynx. From the pharynx the food passes into a short tube, the esophagus, into which small glands forming a calcareous fluid pour their secretion. The esophagus opens into a large, thin-walled chamber, termed the crop, which acts as a reservoir for the food before it is passed into the next region of the food canal termed the gizzard. This is a thick-walled chamber with horny walls, leading into the intestine, which is a long, straight tube running to the end of the body, where it opens to the exterior at the anus. The intestine is covered with a layer of yellow cells, and projecting into its cavity from the dorsal wall is a ridge called the typhlosole.

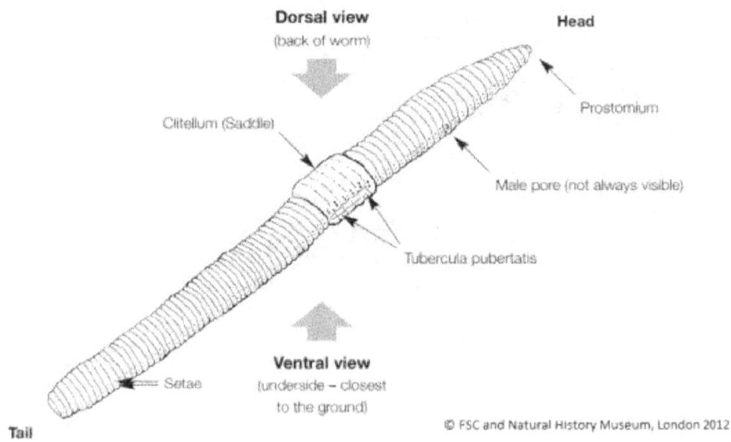

The layer of cells which line the inside of the intestine is called the endoderm and corresponds to the endoderm of hydra. The alimentary canal is supported and held in position by the septa.

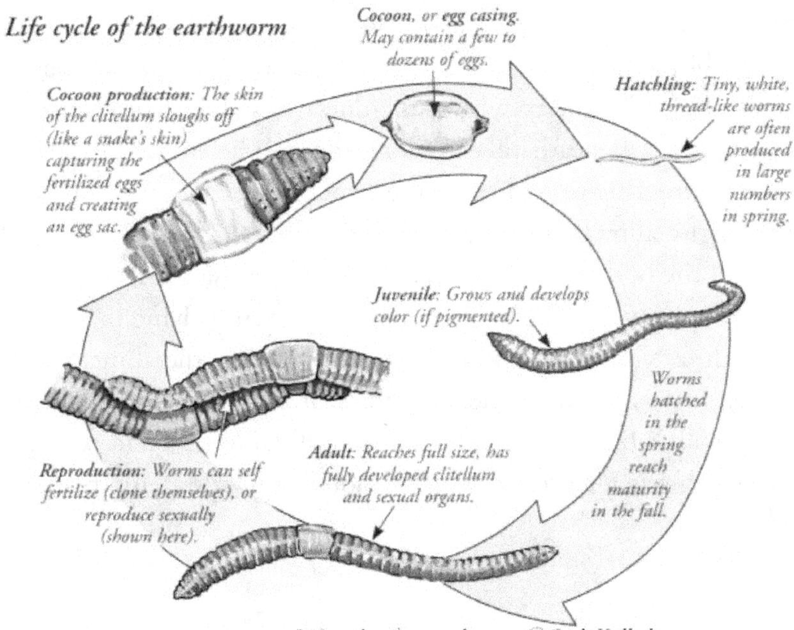

Life cycle of an earthworm © Rick Kollath

The Blood System

The blood of the worm consists of a liquid, plasma, which contains small colorless amoeboid cells. The plasma is colored red by the pigment hemoglobin which is dissolved in it. The blood is carried in a complex system of vessels, one of which, the dorsal vessel, can be seen running as a red streak beneath the ski along the middle of the dorsal surface of the worm. This vessel lies on the upper side of the alimentary canal. Beneath the alimentary canal the ventral vessel runs along the whole length of the body and is connected with the dorsal vessel in the region of the esophagus by five pairs of contractile vessels lying in segments 7 to 11. Three smaller longitudinal vessels lie below the alimentary canal in close relation with the nerve cord. From the five main longitudinal vessels smaller branches ramify to the internal organs and to the body wall, which has a close network of blood capillaries lying beneath the epidermis.

The Excretory Organs

The excretory organs of the worm are called nephridia and occur in pairs in the coelom of each segment except the last and the first three. Each nephridium is a long tube coiled in several loops and opening into the coelom at one end by a funnel-shaped opening termed the nephrostome. The nephrostome does not lie in the same segment as the nephridium but passes through the septum into the segment in front. The nephrostome and part of the nephridium is lined with cilia, and the loops of the nephridium are surrounded by a network of blood vessels.

The Nervous System

Lying in the coelom beneath the alimentary canal is a slender white thread termed the ventral nerve cord. Branch nerves are given off from the nerve cord which pass to the skin and internal organs, and in each segment the cord shows a slight swelling termed a ganglion. On the upper side of the buccal cavity are a pair of small white rounded bodies which are united in the middle line. These are the cerebral ganglia which form the brain of the worm and give off many small nerves to the prostomium. From the cerebral ganglia two bands of nervous tissue pass one on either side of the pharynx to join the anterior end of the ventral nerve cord, which lies beneath the pharynx.

The earthworm has no special sense organs and cannot see or hear. However, the front end of its body is sensitive to bright light, which the worm avoids. It has a well-developed sense of touch and is able to perceive vibrations in the soil. The worm has also a sense of smell, for it can recognize the presence of such food as fat and meat. The sense cells which perceive these various stimuli are scattered in the epidermis, and each has a small sense hair which projects through the cuticle. The bases of the sense cells are connected with the ventral cord or cerebral ganglia by nerves.

Respiration

There are no special organs of respiration; the worm respires through its thin damp skin.

The Reproduction of the Earthworm

The earthworm is hermaphrodite and the male and female reproductive organs occur in segments 9 to 15. These are a pair of ovaries and two pairs of testes. The eggs formed by the ovaries pass to the exterior by a pair of short tubes, the oviducts, which open on the ventral surface of the 14th segment. During pairing the spermatozoa pass down paired spermiducts or vasa deferentia which open on the ventral side of the 15th segment. Two pairs of rounded sacs, the spermathecae, open to the exterior in the grooves between the 9th and 10th, and 10th and 11th segments. Although hermaphrodite, a single worm cannot fertilize its own eggs but must pair with another worm in order that fertilization may take place. The spermathecae store sperm received from the other worm during coition, for self-fertilization does not occur since the structure of the animal makes this impossible.

Two animals lie in opposite directions and secretion from their clitella fastens them together. Their hinder ends lie each in its burrow, so that the animals can retreat if disturbed. Each animal then receives the sperm which are stored in the spermathecae. The animals separate and for the next week proceed to produce cocoons containing eggs and sperm.

During this process the clitellum secretes a thick band of substance and the animal then proceeds to slip backwards through it. Eggs are shed into it through the apertures in segment 14, whilst later, little packets of sperm are shed when the cocoon passes segments 9 and 10.

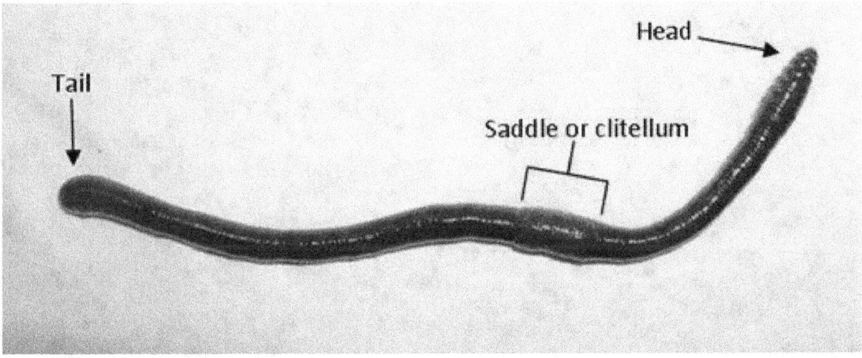

Epidermal cells also secrete a milky fluid into it. Eventually the animal slips its head out of the cocoon which, being elastic, closes up. Its front part is flattish whereas the other end is tapering, so that it is pear-shaped. The cocoon is usually left in a place where there is decayed vegetation. As it dries it becomes darker, tougher and somewhat horny. Conjugation between the gametes now occurs and the zygotes, nourished by the milky fluid, develop into young like the parents. Sometimes three or four small worms will make their way out of a cocoon, but more commonly one develops at the expense of the others.

Worms have considerable power of repairing injuries, for after a worm has been cleanly cut into head and tail parts by a spade, each half usually regrows the missing part so that two worms result.

The Cockroach

There are well over 1000 species of cockroach in the world and most are found in the tropics. We are most likely to come across two of the larger domestic species, i.e. the Oriental Cockroach and the American Cockroach. The former can be distinguished from the latter by its darker color and the fact that the female has no wings.

The body of a mature cockroach is about 1 ½ inches long and is divided into three regions: the head, thorax and abdomen. The head is joined by a narrow neck to the thorax, which consists of three segments and bears the legs and wings. The abdomen is longer than the thorax and is made up of ten segments. The whole body is covered by a dark brown cuticle, consisting of a substance called chitin, which is thick, tough and horny, except at the joints, where it is thin, soft and flexible. Each segment of the body is covered by an upper plate called the tergum, and a lower plate, the sternum.

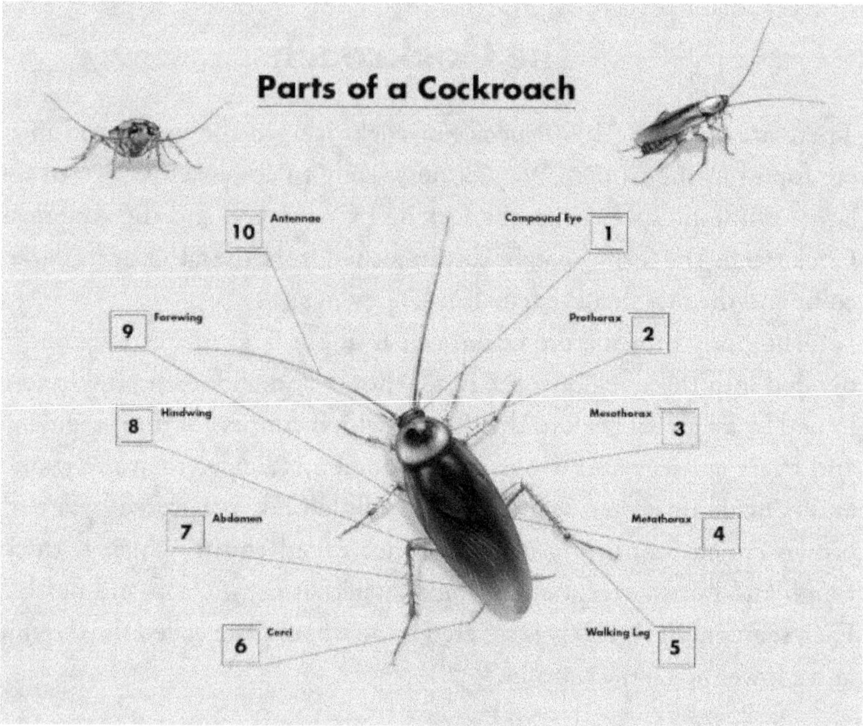

The head is oval in shape and is attached to the body by the short and slender neck. On each side of the head is a large kidney-shaped compound eye. A compound eye consists of thousands of minute eyes or ommatidia packed closely together, and the transparent cuticle which forms the surface of the compound eye is divided into thousands of small polygonal areas or facets. Each facet represents the outer surface of an ommatidium. Projecting beneath each eye is a long, whip-like feeler or antenna, which is a many-jointed structure bearing organs sensitive to touch, smell and vibration. The cockroach is constantly feeling the ground with its antennae, and it also waves them above its head in order to test the air.

From the lower front edge of the head a flap called the labrum or upper lip hangs in front of the mouth, at the sides of the mouth there are three pairs of jaws which work from side to side.

The first pair of jaws are called the mandibles. Each has a strongly toothed inner edge used for biting. Behind the mandibles lie the second pair of jaws called the first maxillae. Each first maxilla is a jointed, forked structure with an inner and outer fork. The base of a first maxilla consists of two joints, and the outer prong is a five-jointed feeler or pulp. The inner prong forms a biting blade. The blades of the two maxilla work against each other and break up the food masticated by the mandibles into even finer particles The third pair of jaws are also forked, but the bases of each second maxilla are fused together to form the labium or lower lip. The outer prong of the fork is a three-jointed feeler or pulp, while the other prong is sub-divided into an inner pulp and an inner feeding blade.

In the cockroach, as in all insects, the thorax consists of three segments, called the prothorax, mesothorax and metathorax. The tergum of the prothorax is larger than that of the other two segments, and overlaps so as to hide the neck in front and part of the mesothorax behind. The mesothorax and metathorax each bear a pair of wings. The first pair of wings are thick chitinous plates, which extend backwards so as to cover two-thirds of the abdomen. They are called elytra or wing cases, and protect the second pair of wings when these are not in use. When the insect is flying, the elytra are held out stiffly at right angles to the body. The second pair of wings and delicate folds of chitin, strengthened by thicker ridges or veins, and when at rest are folded fan-wise beneath the elytra. In the female of Blatta orientalis these wings are reduced to stumps. A pair of legs are attached to the under surface of each thoracic segment. Each leg consists of a number of joints, and ends in a pair of claws.

Internal Structure of a Cockroach

Only eight of the ten segments of the abdomen are visible, since the 8th and 9th are overlapped by the tergum of the 7th segment. The last tergum is grooved along its posterior edge, and the anus opens beneath it. At the sides of this segment are a pair of jointed, spindle-shaped structures, the cerci, which are supposed to be sensory in function. In the male the 9th segment bears a pair of short styles, and in the female the 7th segment has a boat-shaped sternum which projects backwards so as to form a pouch in which the egg case is formed.

Madagascar Hissing Cockroach

If the terga are carefully cut away from the back along the thorax and abdomen of the cockroach, a mass of loose white tissue is seen. This is the fat body which largely fills the space between the alimentary canal and the body wall. This space is not a coelom and so cannot be compared with the body cavity of the worm or a vertebrate, but is a blood space in which the organs lie bathed in blood. Lying just below the terga and extending along the middle of the thorax and abdomen is the narrow contractile heart, which by its action forces blood to circulate through the sinuses. Such a blood system is said to be an open system, whereas the arrangement of narrow blood vessels possessed by the worm and by vertebrates is a closed system. The blood is colorless and contains small amoeboid cells.

Digestion is the concern of the alimentary canal which lies beneath the fat body and is partly enveloped in it. Food that has been broken by the mandibles is softened by a digestive juice secreted by a pair of salivary glands lying each side of the crop, but whose ducts join together to open on to the mouth by one common duct.

The mouth opens into a narrow tube, the esophagus, which is followed by a large pear-shaped crop and the gizzard, which has thick walls and six horny teeth which grind up the food. Behind the teeth are a number of bristles which strain the food before it passes into the next part of the alimentary canal, which is a short narrow tube termed the mesenteron or mid-gut. The mid-gut is followed by a longer tube, the colon, the end of which is enlarged to form the short rectum. The rectum opens to the exterior by the anus at the hind end of the abdomen. The most notable feature is that only the mid-gut is lined by endoderm and conveys digested food through its walls to the surrounding blood. All the other parts are lined by ectoderm.

At the beginning of the colon the numerous fine, thread-like Malpighian tubules are attached. These, interlaced with the lobes of the fat body, form the excretory system.

The respiratory system consists of a series of air tubes, the larger ones opening to the surface at pores along the sides of the body. Their branches divide into tracheoles which penetrate every tissue. All parts are prevented from collapsing by a spiral thickening of chitin.

The nervous system of the cockroach is constructed on a similar plan to that of a worm. In the head there is a large supra-esophageal ganglion or brain, from which nerves pass to the eyes and antennae. A ventral nerve cord runs along the body wall beneath the alimentary canal, and consists of a sub-esophageal ganglion and a chain of nine ganglia united by double connectives. The brain is joined to the ventral nerve cord by a collar which passes round the esophagus.

The male cockroach has a pair of testes which lie in the abdomen hidden by the fat body. They are only functional in the young cockroach, and afterwards gradually disappear.

In the female there are two ovaries, each of which consists of eight tubes which have a beaded appearance due to the enclosed eggs. After fertilization the eggs are enclosed in a chitinous egg capsule which is formed from the secretion of certain glands in the abdomen of the female.

Life History and Habits

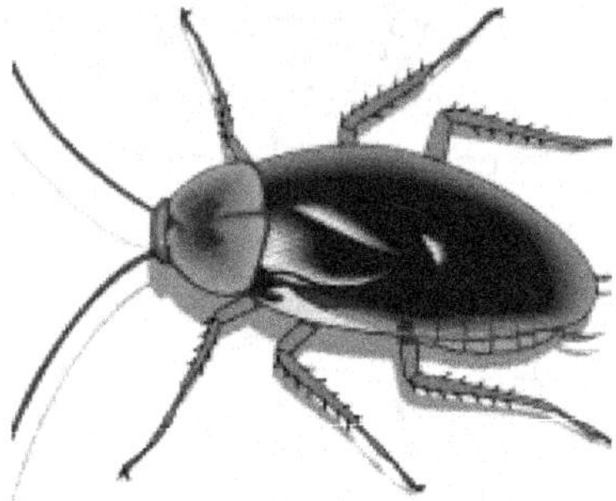

The domesticated species of cockroach are omnivorous. They are particularly fond of sweet and starchy things, but in the tropics wherever there is food there will one find cockroaches. In the absence of food or dead insects they will feed on clothing, paper, books and even shoes. Thus they are the enemy of the householder the world over.

The cockroach has two noticeable characteristics. Firstly it shuns light and haunts cupboards, drawers, and darkened rooms. Secondly its body is so flattened that it can hide in narrow crevices, behind shut drawers, in fact in places where it is difficult to find and destroy.

As it moves about and feeds in the dark its large eyes cannot be of much use for vision. Instead it relies upon its long antennae for feeling and smelling. Cockroaches can frequently be seen cleaning these antennae by passing them through their mouthparts. Wings likewise are brushed by the legs.

The eggs are laid in a horny purse-like egg case which is formed beneath the abdomen of the female, and may be carried about by her for a while before she deposits it and cements it in a crevice or to a surface. The egg case of the Oriental cockroach is made of chitin and contains sixteen eggs arranged in two parallel rows of eight. After a while these eggs hatch into nymphs, which are identical with adult cockroaches except that they are very small, wingless and almost without color. The nymphs feed, become darker and soon outgrow their horny skin. This skin splits and is cast, the insect first producing beneath it another skin that fits its expanding size and which soon hardens. This happens six or seven times until the cockroach is fully grown and wings have been fully formed. Once development is complete, skin-casting ceases. This process is followed by all growing insects that do not go through larval and pupal stages. The method described for the cockroach is called incomplete metamorphosis, while that of the butterfly is complete metamorphosis.

With their habit of hiding in boxes it is easy to see how cockroaches can unwittingly be transferred from place to place by man. Most ships swarm with cockroaches and this explains the presence of domestic species throughout the world.

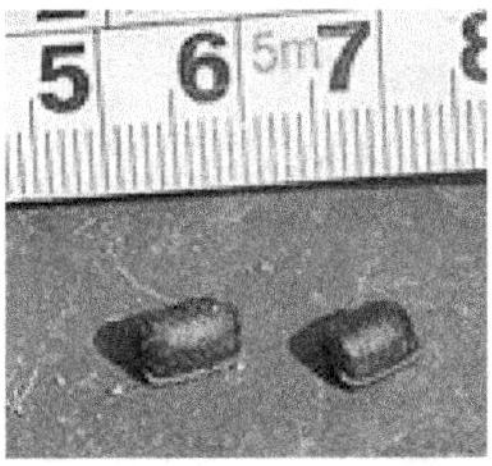

Also by nbillett

Discovering Jamaica
A Journey Through Jamaica's Animal Kingdom
Animals of Jamaica
Jamaican Animals
Sites, Towns and Villages of Jamaica: Volume I
Sites, Towns and Villages of Jamaica: Volume II
The Vampire Connection
Historic Plantations of Jamaica
Botanic Gardens, Parks and Scenic Attractions of Jamaica
Historic Religious Structures, Legends and Legendary Beings of Jamaica
Birds and Rivers of Jamaica
Geographical and Geological Features of Jamaica
The Earthworm and The Cockroach

www.ingramcontent.com/pod-product-compliance
Lightning Source LLC
LaVergne TN
LVHW022033200926
848481LV00007B/2668